All about GARBAGE COLLECTORS

Brianna Kaiser

Lerner Publications ◆ Minneapolis

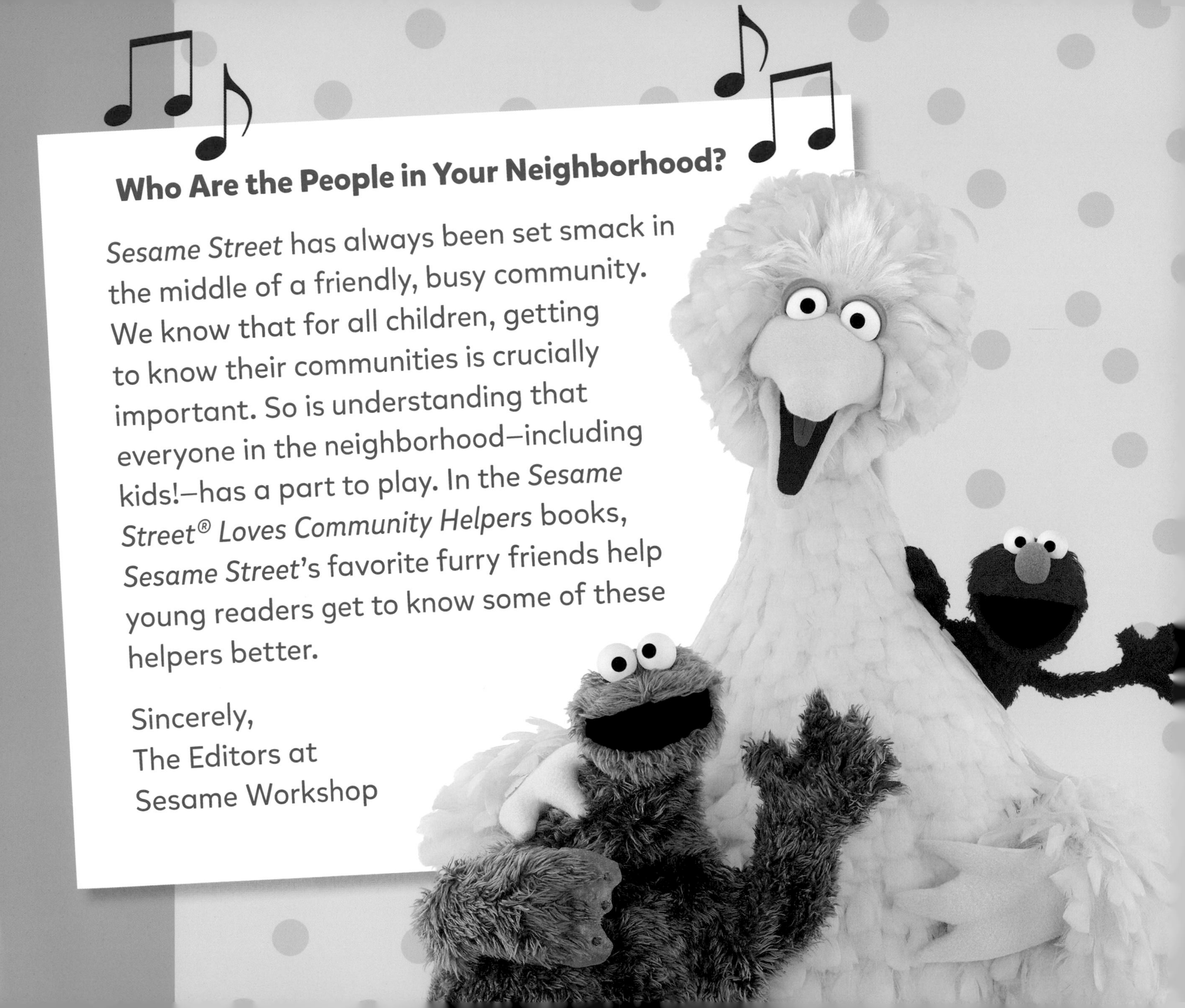

Who Are the People in Your Neighborhood?

Sesame Street has always been set smack in the middle of a friendly, busy community. We know that for all children, getting to know their communities is crucially important. So is understanding that everyone in the neighborhood—including kids!—has a part to play. In the *Sesame Street® Loves Community Helpers* books, *Sesame Street*'s favorite furry friends help young readers get to know some of these helpers better.

Sincerely,
The Editors at
Sesame Workshop

Table of Contents

We Love Garbage Collectors!

Learn about Garbage Collectors

Garbage collectors make the community a better place. They work hard to keep places clean.

I help my parents keep our home and community clean by collecting trash.

Garbage collectors pick up trash, recycling, and yard waste.

They all go in separate bins.

Garbage collectors work in many places. They take away trash and recycling from homes, parks, and businesses.

I wave to garbage collectors when they come to my home.

They drive garbage trucks and recycling trucks. Some trucks have an arm that picks up the bins.

Don't take my bin!

If the truck doesn't have an arm, a collector empties the bins into the trucks.

I am strong
like garbage
collectors!

Garbage collectors sometimes wear bright-colored shirts or vests. This helps other drivers see them.

Garbage workers wear ear and eye protection!

They also wear gloves.

Gloves protect their hands and keep them clean.

Garbage collectors wear clothes to protect them from the weather.

They work outside when it's warm and cold.
They work in sunshine, rain, and snow.

When their trash truck is full, they drive to a center where they empty the trash. Some trash is taken to a landfill.

There must be lots of trash at the center!

Recycling goes to its own station. It is sorted.

Yard waste goes to a compost area.

Picking up trash helps keep us, animals, and the planet healthy.

I say thank you to my garbage collector.

Thank you, Garbage Collectors!

Dear Garbage Collectors,

Thank you for everything you do. I love trash, but thank you for keeping places clean.

Your friend,

Oscar

Now it's your turn! Write a thank-you note to your garbage collectors.

Picture Glossary

bins: containers that hold trash or recycling

community: a place where people live and work

compost: a type of fertilizer made from rotting plants

recycling: turning trash into something new and beautiful

Read More

Harasymiw, Martin. *Garbage Trucks*. New York: Gareth Stevens, 2022.

Lindeen, Mary. *Reduce, Reuse, and Recycle, Oscar!* Minneapolis: Lerner Publications, 2020.

Murray, Julie. *Garbage Collectors*. Minneapolis: Abdo Kids Junior, 2019.

Index

Photo Acknowledgments

Image credits: Blend Images - Don Mason/Getty Images, pp. 5, 8; kzenon/Getty Images, p. 6; SDI Productions/Getty Images, pp. 7, 26; SAV_____A/Getty Images, p. 9; brittak/Getty Images, p. 10; tylim/Getty Images, p. 11; M2020/Shutterstock.com, p. 12; Paul Vasarhelyi/Shutterstock.com, p. 13; James Hardy/Getty Images, p. 14; Sergio Shumoff/Shutterstock.com, p. 15; Don Mason/Getty Images, pp. 16, 17; sutiporn somnam/Getty Images, p. 18; PeopleImages/Getty Images, pp. 19, 20, 29; picture alliance/Getty Images, p. 21; Pramote Polyamate/EyeEm/Getty Images, p. 22; vm/Getty Images, p. 23; Mikael Vaisanen/Getty Images, pp. 24, 30; BuildPix/Construction Photography/Avalon/Getty Images, pp. 25, 30; SDI Productions/Getty Images, p. 26; 4x6/Getty Images, p. 27; TOBIAS SCHWARZ/AFP/Getty Images, p. 30.

Cover image: Don Mason/Getty Images.

Lerner Publications Company
An imprint of Lerner Publishing Group, Inc.
241 First Avenue North
Minneapolis, MN 55401 USA

For reading levels and more information, look up this title at www.lernerbooks.com.

Main body text set in Mikado Medium.
Typeface provided by HVD Fonts.

Designer: Mary Ross
Lerner team: Martha Kranes, Sue Marquis

Library of Congress Cataloging-in-Publication Data

Names: Kaiser, Brianna, 1996- author.
Title: All about garbage collectors / Brianna Kaiser.
Description: Minneapolis : Lerner Publications , [2023] | Series: Sesame Street loves community helpers | Includes bibliographical references and index. | Audience: Ages 4–8 | Audience: Grades K-1 | Summary: "Oscar loves trash and garbage collectors! Garbage collectors pick up garbage, recycling, and yard waste. Learn more about these community helpers!"– Provided by publisher.
Identifiers: LCCN 2021040275 (print) | LCCN 2021040276 (ebook) | ISBN 9781728456133 (library binding) | ISBN 9781728462110 (ebook)
Subjects: LCSH: Sanitation workers—Juvenile literature. | Refuse collectors—Juvenile literature. | Refuse collection—Juvenile literature. | Refuse and refuse disposal—Juvenile literature.
Classification: LCC HD8039.S257 K35 2022 (print) | LCC HD8039.S257 (ebook) | DDC 628.4/42023–dc23

LC record available at https://lccn.loc.gov/2021040275
LC ebook record available at https://lccn.loc.gov/2021040276

Manufactured in the United States of America
1-50683-50102-1/21/2022